I0059658

Staying Safe while Conducting Hands-On Science

Safety Guidelines for the Parents or Adults Conducting Hands-on Activities with Children

By

Frankie Wood-Black, Ph.D.

First Edition
2013

Staying Safe while Conducting Hands-On Science: Safety Guidelines for Parents or Other Adults Conducting Hands-on Science Activities with Children

Written By: Frankie Wood-Black, Ph.D., REM, MBA
Copyright 2013 Sophic Pursuits, Inc.

All rights reserved. No part of the book may be used or reproduced in any manner without written permission of the author/publisher.

Publisher's Cataloging-in-Publication data
Wood-Black, Frankie, 1963-
      Staying Safe while Conducting Hands-On Science:
      Safety Guidelines for Parents or Other Adults
      Conducting Hands-on Science Activities with Children

ISBN 978-1-940843-01-8
1. Science – Study and Teaching. 2. Education – Activity Programs. 3. Safety Education.

Acknowledgements

American Chemical Society
K-12 Education Materials
&
The Joint Board - Counsel
Committee on Safety

Division of Chemical Health and Safety
of the American Chemical Society

As much of science is built upon the knowledge of others, it is fitting to recognize those who have gone before. The American Chemical Society has been a leader in chemical information and the development of educational resources. Thus, it is fitting to acknowledge this body of work.

Dedication

I would like to thank my family and Harry Elston, Ph.D. for helping me get this endeavour out the door. Without their encouragement and advice this book would have never made it to press.

Disclaimer

While the author has used accepted references and has consulted with individuals with experience in health and safety related matters, no individual safety reference will be able to address every possible hazard and/or situation. Additionally, each individual learning situation will be different and the potential hazards associated with specific hands-on activities may or may not have been addressed by the developer of the activity. Thus, this document is intended to be used as a starting point.

The author is not making any claims, warranties, or guarantees related to the sufficiency of the information contained in this document. As the specific situation and/or environment cannot be anticipated by the author, this document in no way outlines all the necessary warnings and precautionary measures which may need to be taken prior to conducting a hands-on activity. Nor, does this document meet or comply with the requirements of any safety code or regulation.

Users of this document should consult specific safety information provided with the materials, instructions, and/or other reference materials to determine hazards associated with each individual activity. Warnings and precautions should be reviewed and evaluated prior to conducting any hands-on activity. All hands-on activities involving children should be conducted under direct supervision.

Contents

Introduction

Science is best learned through discovery. Demonstrations and hands-on activities are therefore an essential and fun aspect of the learning experience. It is through experimentation that the excitement of discovery leads to a better understanding of the world around us. This excitement leads to the development of skills that will be used throughout life. Skills that can be learned through hands-on activities include but are not limited to:

- Observation
- Documentation
- Communication
- Critical Thinking
- Scientific Method

There are many sources of hands-on activities. Your local library probably has shelves of hands-on activities related to science. The local book store will have a number of books that claim to have hosts of science experiments to do at home, or in the kitchen. One quick search on the internet can lead to numerous activities and YouTube™ videos of experiments. While this is wonderful for teachers, parents and students, there are risks associated with these hands-on activities. There is the question of how safe are the activities?

Risk versus Hazard

In the world of the safety and environmental professional, the words risk and hazard are used on a daily basis. And, most of the time it may seem that the words are used interchangeably. This may be due to the behavioral response by people. Fundamentally, the words are different – yet how it is how people respond to the specific situations that tend to cause the confusion.

Hazards are defined as a source of potential adverse effects on someone or something. The adverse effect can result in physical damage or an adverse health effect. A hazard is typically a thing or situation. For example a wet floor which could lead to a slip or fall; a sharp edge which could lead to cuts; contaminated atmospheres which could lead to a harmful health effects; or prolonged exposure to heavy metals which could lead to poisonings.

Risk is the chance or probability that an adverse effect will happen in the presence of a hazard. A sharp edge does not always result in a cut. A wet floor does not always result in a slip and a slip does not always result in a fall.

Just because a hazard is present does not mean that the adverse effect will happen. It is the behavior of individuals and the specific conditions associated with the hazard that result in the adverse effect. Risk is associated with perception, how an individual views the hazard; frequency, the number of times a person is exposed to the hazard; likelihood, how easy it is for the hazard to result in an adverse effect; and magnitude, the severity of the adverse effect.

Most educators, scientists, and activity developers are concerned about the materials that they present. No one wants anyone to be hurt while they are exploring the wonders of science. No one wants to inadvertently expose a child or an adult to a hazard. Yet, many of these materials fail to describe basic safety or warn of the hazards associated with the particular activity. Many will indicate that the activity needs to be completed with direct adult supervision, but fail to taken into account that the adult may or may not have the background to anticipate some of the risks and/or potential unintended consequences.

This book has been designed with the homeschool or cooperative school instructor in mind. Additionally, this book can be used by any parent that may wish to provide supplemental or enhancing activities that expand the information being presented in a typical classroom or to just promote their child's interest in science. The idea behind this book is to provide basic safety guidelines and precautions so that the hands-on activity remains fun and there is not an unintended learning experience.

While basic safety guidelines and precautions are presented, there is no way that a single resource can anticipate all the potential conditions and situations that may occur. Thus, this book can in no way be an exhaustive resource. It is intended that the information presented here can:

- Outline basic safety guidelines
- Provide a method for reviewing the activity prior to conducting it to determine potential hazards
- Provide suggestions for resources or where to access safety information
- Provide a code of conduct

It is hoped that by providing safety guidelines that the learning experience can be enhanced. Hands-on activities are a way that the thrill of science can be communicated. Without these hands-on experiences, science loses its wonder. Yet, these activities need to be safe and should not present an exposure or hazard.

Safety is something that happens between your ears, not something you hold in your hands.

Jeff Cooper,
United States Marine

Basic Safety

Safety starts with you and your attitude toward preventing accidents. Safety is about behavior, the ability to recognize hazards, and being prepared to handle an adverse situation. A hazard is a source of a potential adverse effect or consequence. A dedication to minimizing hazards and thinking prior to acting is key to preventing accidents when participating in any activity. As individuals, we tend to overlook hazards because of our pre-occupation with other demands on our attention. Or, we become complacent and accept a certain level of risk because of a sense of familiarity or lack of understanding. Risk is the probability that the adverse effect or consequence will occur.

Just think for a brief moment about the normal activities you perform every day. You get up and generally go straight to the bathroom. This behavior has certain risks as according a 2011 publication of data from the analysis of emergency room visits in 2008 by the Center for Disease Control and Prevention. According to this analysis approximately 235,000 people each year over the age of 15 visit emergency rooms due to an injury suffered in the bathroom. Yet, this is an activity that most people don't perceive as risky due the familarity of the actions.

Risks typically fall into three categories – unknown, risks we choose, and risks placed upon us. Looking at these more closely, risks that are placed upon us are those that we have little or no control over.

For example, you live in a community that is near an ocean. You and your family may be at risk from dangers associated with flooding, hurricanes or tsunamis. Other types of risk that individuals have limited control over are risks associated with businesses in your community or risks associated with the actions of others.

Risks that we choose might include risks associated with driving or hobbies. You choose a certain level of risk if you opt to play football rather than reading a book. These are risks that you may have a certain level of comfort with due to your understanding or knowledge. How frequently you encounter the risk may impacts your perception.

The final category – unknown risk – is the toughest to evaluate. For most, it is an unknown risk because we fail to notice or observe the risk. We aren't looking for it; therefore, we don't see it.

Why should we be concerned about risk?

Risky behavior leads to accidents. To illustrate this point, let's look at an activity that routinely takes place in the kitchen – chopping or cutting vegetables. Most of the time this activity occurs without any adverse outcome, the vegetables are chopped and are ready for use. However, sometimes your attention may be averted or the knife is not as sharp as it should be or your hands are wet, and the result is that you nick your finger. The nick is nothing serious but it reminds you that you were not paying adequate attention to the activity. Occasionally but generally less frequently, the injury may be more severe and may even result in an emergency room visit for stitches. This is a great example of the hierarchy of injury. (See Figure)

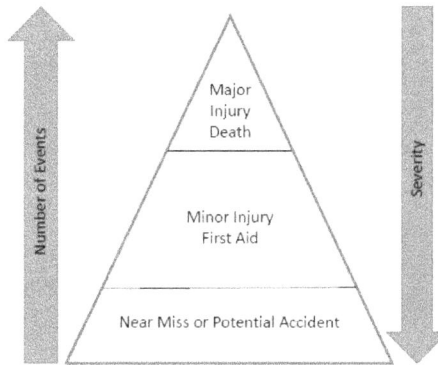

Figure 2.1 – Hierarch of Injury. In this graphic, the severity of the injury is highest at the top, while the frequency of injury is highest at the bottom.

Here you see that most of the time our risky behavior does not result in a bad outcome. Yet, if we persist, we may see an increase as we move toward the top of the pyramid in the severity of the potential injury. The numbers are great on the bottom of the pyramid and decrease (at least it is hoped) as we move toward the top. This is also referred to as the "Safety Iceberg." You can imagine if the pyramid were floating in the ocean, only the top of the pyramid would be showing.

Applying Safety to Hands-on Activities at Home

As a parent in either a homeschool or extracurricular activity situation, your home will probably be the location for many of your experiments. If you are in a cooperative school situation, someone's home, or other location may be chosen. This means that you are going to have to be aware of the potential risks associated with your location and the activity that you choose to conduct. The idea is to minimize the risks.

General Household Safety

The first step in this process is to look at your home or planned location where you are going to perform the experiments from a general common sense safety perspective. Your first step might be to think about all of those National Prevention or Preparedness Week public service announcements. Is your home or school prepared for common hazards, i.e. the typical household or building hazards? What should every location have in the way of safety equipment? Your typical home, school, or building needs to have the following fire and safety equipment according to the National Fire Protection Association:

- Fire extinguishers
- Smoke alarms
- Carbon monoxide detectors

In addition to fire protection, you should have an emergency first aid kit equipped to handle basic medical and/or injury emergencies. The National Safety Council (www.nsc.org) and the American Red Cross (www.redcross.org) have recommendations for what should be included in a household first aid kit as well as general emergency kits.

Just having the equipment is not all that is required. The equipment needs to be in workable condition. You also be able to answer the following questions:

- Where is the equipment located?
- What are the warnings produced by the detectors and alarms?
- What are the actions to be taken in the event or an alarm?
- How the equipment is used?
- Who to contact in the event of an emergency and type of emergency?

Not only do you need to know this information, but everyone who will be participating in your hands-on activity will need to know where the essential equipment is, what to do in case there is a mishap, and how to go for help.

The figure provides a listing of emergency numbers that you should post by your phone, or have pre-programmed into your cell phone for your area. Depending upon your cell phone, tablet, or computer, you may want to have an emergency application at your fingertips. The American Red Cross has an application that can be downloaded for Andriod™ or iPad™.

Emergency Phone Numbers

General Medical or Police Emergency	**911**
Poison Control	**1-800-222-1222**

Other Emergency Contacts

Police	_____
Fire Department	_____
Emergency Room	_____
Primary Doctor	_____

Other Contacts

_____	_____
_____	_____
_____	_____
_____	_____

Figure 3.1 – Emergency Numbers

Recall, the National Fire Protection Association recommends that you have an evacuation plan and a meeting location in the event of a fire. Everyone that may be participating in your activity needs to know the evacuation plan and meeting location. You may even want to have a map to use as a discussion tool.

Basic Safety Considerations for Hands-On Activities

Prior to starting any activity whether it be a science experiment or starting to cook dinner, safety should be an initial consideration. Years ago DuPont, known for its workplace safety record, started the "Take Two™" program to enhance worker safety. This program spread throughout industry and ultimately was carried home by workers so that the "Take Two™" concept was applied to everyday tasks. The idea behind the program was simple – stop and take two seconds and if appropriate two minutes to think about the task that was about to be performed. By doing this simple act of stopping and thinking about the task to be done, it put the person about to perform the activity into a mind frame that allowed them to assess the situation.

Yet, as in the workplace; you need to review the activity about to be performed. It may have been previously reviewed and evaluated, but no previous review can anticipate your specific situation, location or circumstances. In the case of our home experiments or science activities, we are relying on the author or the presenter of the material to have thought about the potential unintended outcomes. The developers should made choices to minimize the risk associated with the activity, but your specifics still need to be considered. We want the activity to be safe. But, in order for you, the user, to be safe; you need to understand the different levels of safety that should be taken during the development of the experiment or activity.

Health and safety professionals call these levels of safety, or the hierarchy of protection. This hierarchy can be used to prevent unnecessary exposures and to minimize risk. Over the past two decades, many consumers have seen this hierarchy in practice without even realizing it, or if they did realize something was different it was usually to complain about a new guard or safety cutoff switch. The figure presents a simplified version of this hierarchy.

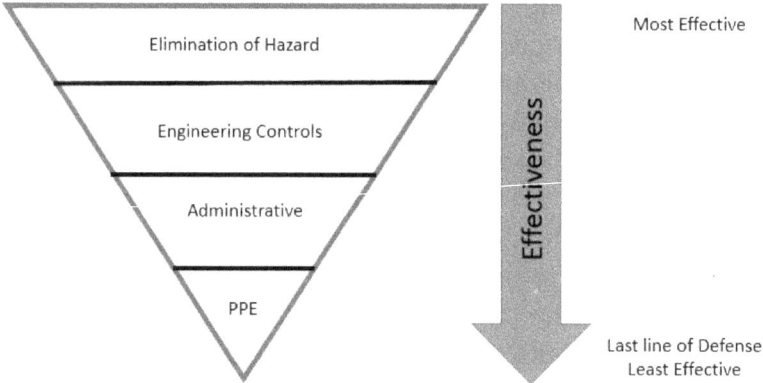

Figure 4.1 – Hierarchy of Safety

The preference is to eliminate the hazard. This is the elimination or substitution phase. For example: you encounter an old fan at a garage sale. It looks cool, but really look at the fan is there a hazard? There are two potential hazards immediately obvious. The protective cage around the blades is not nearly as protective as we have come to expect from a modern fan. And, the electrical cord is probably not in the best condition. It is likely that the insulation around the wires is compromised.

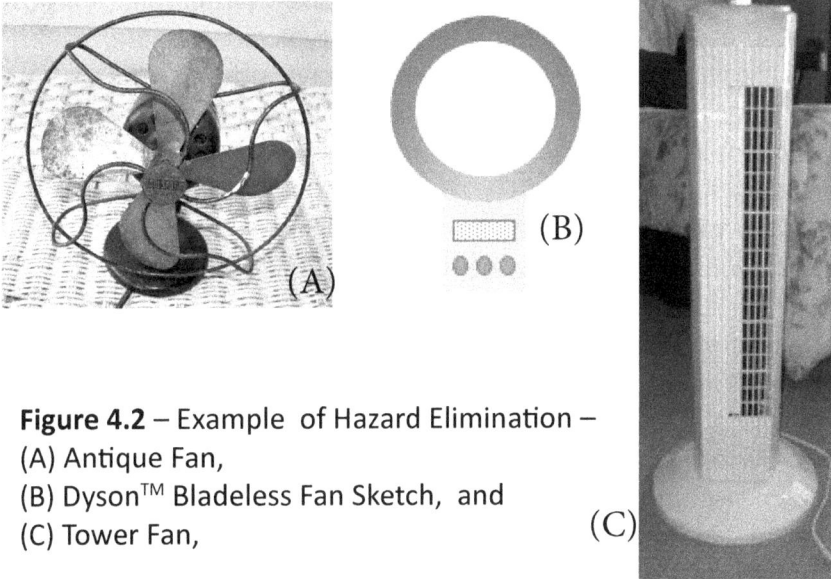

Figure 4.2 – Example of Hazard Elimination –
(A) Antique Fan,
(B) Dyson™ Bladeless Fan Sketch, and
(C) Tower Fan,

It is likely that the insulation around the wires is compromised and it probably is not grounded. Other electrical hazards may also be present. In some modern fans, the hazard from the rotating fan blades has been eliminated completely by using a different means of moving the air, see the Dyson™ bladeless fan or in tower fans.

The second aspect of elimination is substitution. For the developers of hands-on activities, substitution is one of the methods that is used extensively.

Let's look at how individuals create a science activity for someone outside of the traditional classroom or laboratory space. Generally, it starts with selecting a concept to demonstrate. This could be a chemical reaction, or a physical concept like momentum. In most cases, the developer is familiar with a typical laboratory experiment that has been used in colleges or universities to demonstrate the concept. Then, they look at how it might be adapted to be used by individuals at home. This, of course, involves several different considerations not just safety. The developer needs to consider the access to the materials needed to perform the experiment as well as the equipment that may be needed.

Let's take for example a simple chemical reaction experiment. In this experiment, we would like to demonstrate the formation of a precipitate, i.e. mixing two liquids in which two different ionic compounds like salt are dissolved and the mixing results in a solid. (See image) In a college classroom demonstration, the lecturer is likely to use an aqueous solution of lead nitrate (lead nitrate dissolved in water) and an aqueous solution of potassium iodide (potassium iodide dissolved in water). The reason these two solutions are used is because the precipitate formed is bright yellow in color and makes for a good visual demonstration. While potassium iodide may be found at your local drug store, the lead nitrate is not as readily available. Additionally, the lead in the lead nitrate poses a potential health hazard for children. So, a substitution needs to be made. The same concept can be demonstrated using materials that are more familiar and are potentially in the household already, Epsom salt (magnesium sulfate) and washing soda (sodium carbonate). Thus, the hazard associated with this activity has been reduced by making

Figure 4.3 – Precipitation Experiment – The top photo shows the experimental set-up. The middle photo shows the two ionic liquids prior to mixing. The bottom photo shows the particle formation or the precipitate. The area is free from clutter, and the person doing the mixing is wearing a glove. (See glove discussion)

a chemical substitution. Similarly, other experiments can be modified to reduce the potential hazards.

Yet, it needs to be understood that just because the hazard may be eliminated, reduced or minimized; a hazard may still be present, or a new hazard may have been introduced. Further actions are required to continue to improve the safety of the overall activity. This leads to the next step in the safety hierarchy, engineering controls.

Let's look at the antique fan again. There is an engineering control in place, although not a very effective one, the guard around the fan blades. In typical modern fan, the guard has been redesigned in such a way as to prevent fingers from entering the area where the blades are circulating. We encounter engineering controls every day in the form of guards, safety switches and automatic cutoffs. Breaker boxes are a form of electrical engineering controls for your home. Other common examples include: ground fault interrupters, requiring the brake to be engaged before you can turn the ignition in your car, forcing your car to be in park prior to removing the key, cutoff switches on lawn mowers, as well as guards on hand tools. All of these features are designed into the equipment to prevent common types of accidents, fires, or injuries. While most experiments or hands-on activities may not directly involve the application of an engineering control, the tools or equipment that is used may have one of these controls in place.

Never remove or disengage
an engineering control.

For hands-on activities the final two lines of protection are administrative controls and personal protective equipment. In the workplace, administrative controls are the workplace rules. These workplace rules are in place to keep everyone safe by setting various limits. The limits may be associated with who can enter certain areas of the facility, who can perform certain types of work, how long a person can be in a specific area, what types of clothing are required, and what steps must be taken when performing specific tasks. These rules are based on the tasks to be conducted, training that may be required, and general accountability. For example: a person may have to check-in prior to entering a restricted area to ensure that a hazardous activity is not being conducted or barricades have to be put in place when unloading a vehicle. These are all examples of administrative controls. These controls don't necessarily remove the hazard, but they help to prevent unnecessary exposures to the hazard.

For the hands-on activity, an example of these types of controls would include:

- Choosing which tasks need to be performed
    by the adult
- Choosing which child performs what task
- Setting up barriers between the experiment
    and surroundings

Let's re-examine the precipitate example. For this activity, the Epsom salt and the washing soda will need to be weighed in order to make the aqueous solutions. Water will have to be measured and poured into the containers in which the solutions will be made. Mixing of the the solids into the water needs to occur. Finally, the mixing of the two liquids will complete the experiment. If you are doing this experiment with several children, after reviewing the activity, you may select the older or more experienced child for the weighing of the solids and allow the younger less experienced child to measure and pour the water. Under supervision, you may allow all the children a chance to stir the solution. Finally, you as the adult will conduct the last part of the experiment pouring one solution into the other.

It is up to you as the instructor and responsible individual to understand the abilities of each of the participants. While you can let everyone participate to varying degrees, you need to consider the skill level required and the potential hazards associated with each step in the process. How the experiment or activity is to be completed is determined by you, the leader of the activity.

In our precipitate example above, one critical element and the last line of defense was omitted during that discussion. The omission was that of the required personal protective equipment or PPE. PPE is designed to protect the user from potential hazards and is generally specific to the particular hazard anticipated. PPE includes but is not limited to: eye protection, hearing protection, hand protection, protective clothing and specialized shoes. We commonly see people wearing PPE all the time. Construction workers have specialized clothing, hard hats, face and eye protection as well as hearing protection. Similarly, scientists have specialized PPE. The selection of the PPE is dependent on the activity being performed.

So, reviewing the precipitate example, what PPE should be in place? What are the potential hazards? For this experiment, the potential hazards are associated with the solids, i.e. getting the Epsom salts on the hands or in the eyes. Then, there is the liquid which also has the opportunity to be splashed or spilled which may get the material on the hands, or clothes and potentially into the eyes. Thus, eye protection, hand protection and some over-clothing such as an apron or smock should be worn. However, it is essential that the PPE be fitted properly to ensure that it in and of itself does not pose a safety risk.

A Word about Gloves and Hand Safety

- Hand injuries are very common.

- Contamination of food and drink can occur as a result of transfer from the hands.

- Chemicals can be absorbed through the skin.

Every activity must be reviewed to determine if there is a specific hazard that needs to be addressed by the use of gloves. Gloves come in a wide variety of materials to meet diverse needs. Additionally, gloves come in multiple sizes. The wrong glove can create a hazard that may be even more hazardous than the one it is trying to mitigate.

As part of your safety review of the activity, you need to carefully evaluate if gloves are needed and what type of glove should be used. Your chosen activity will be your first source. Other sources of information that should be reviewed to determine whether or not a glove should be used and the recommended type of glove will include the product label or associated product information, and any safety data sheets.

Finally, ensure the glove fits properly.

Basic Personal Protective Equipment (PPE) for Hands-On Activities

Personal protective equipment is that last bit of insurance that is needed to minimize potential injuries associated with conducting hands-on activities. Additionally, it is very cheap insurance. According to a recently published National Institutes of Health study, 2013, an average trip to the emergency room can cost approximately $740 or greater than a month's rent. Compare this with the cost of a pair of child sized safety goggles, $3.00 to $ 7.00. This doesn't even take into account the potential loss of eyesight or an eye, because goggles were not being used. So, what is the minimum personal protective equipment that should be available prior to conducting your hands-on activity?

For most activities, the basic personal protective equipment should include:

• Eye protection – splash goggles for chemical experiments, safety glasses for experiments or activities where particles may become airborne.

• Hand protection – gloves appropriate to the activity need to be present. This may include gloves that are used to minimize scrapes and cuts, or gloves to prevent liquids from coming into contact with the skin.

• Hearing protection – ear plugs or muffs may be required if your activity involves loud noises – either continuous noise or impact noise.

Key things to remember about personal protective equipment (PPE):

> • It can't protect you if it is not worn.
> • It can't protect you if it does not fit.
> * It can't protect you if it is worn incorrectly.
> • You don't want the PPE to become a hazard
>   by itself – watch for loose or damaged items.

In addition - to the PPE - you need to be aware that jewelry and/or long hair may also pose a hazard. Make sure that long hair is tied up so as not to get caught or inadvertently come into contact with the materials being used. Similarly, jewelry should be removed so as not to get caught.

Other PPE may be required for your specific activity. Read and review the activity prior to beginning any experiment to determine the specific PPE that may be required. Make sure that you assess the risks associated with the activity based on the recommended PPE.

Science is a way of thinking much more than it is a body of knowledge.

Carl Sagan

Preparing your "Laboratory" Space

Now that you understand the hierarchy of safety as well as the minimum personal protective equipment, it is time to address how to set up your activity space. The specific activity being performed will usually dictate the where you are likely to set up. Odds are the two most likely locations will be in the kitchen or somewhere outside.

As discussed in the basic considerations, you will need to assess your specific area and make sure that you have ready access to any safety equipment that you might need. This will also include clean-up materials for spills, water, and trash generated during the activity. You need to make sure that your space is free of clutter and/or extra materials that may make it difficult to perform your activity or may become a hazard as your activity progresses. So, let's look at our two locations in more detail.

Kitchen Hands-On Activities

Not only are you trying to conduct the experiment or activity to teach a scientific concept, but you also want to start building good laboratory habits that can be carried on to a high school or college environment. Thus, you need to start setting the boundaries right away.

So, some kitchen science tips and rules to start with:

- Don't contaminate foods or drinks – move them away from your experiment area

- Even though you will be using household or readily available materials – do not eat or drink materials being used or made from your experiment or activity

- Keep your workspace neat and tidy

- Use appropriate PPE

> ## Dangerous Mixtures
>
> Do not mix bleach with ammonia.
>
> Do not mix bleach and acids (vinegar is an acid)
>
> Do not mix bleach with toilet bowl cleaners
>
> Do not use 2 drain cleaners together or one right after the other

The two most likely locations for your specific activity will be a counter top or the kitchen table. You need to make sure that these are free of clutter. You may wish to cover you workspace with some type of protective material depending upon your activity, for example you may wish to use plastic sheeting or newsprint. Try to have only the materials that you will be using as part of your activity in your work space. You don't want to have an inadvertent interaction between the materials you are using and some other chemical and/or a heat source. (See the Dangerous Mixtures sidebar.)

Another consideration is the ergonomics of the activity. Ergonomics is an applied science related to how an individual interacts with a workspace. For our case, home learning activities, we can think of this consideration is how the child, teen or adult is going to interact with the equipment and/or elements of the activity. Reach and height are your first considerations. You will have to take into account how the child is going to be able to access or participate in the activity. Is a stool going to be required? Is the equipment and/or materials to be used sized appropriately?

You may have to do some initial preparation, by putting materials into smaller containers or making sure that the stool to be used is sturdy and free of defects. You may have to adjust or change your work location to accommodate the students who will be participating. Finally, make sure that your safety equipment is readily accessible.

Outdoor Activities

Your chosen activity may be better performed outdoors. Space and the ability to dissipate vapors are a couple of the advantages to moving the activity outside. Hopefully, the vapor issue is not because it is a hazardous, but because the reaction may have a disagreeable smell or generate smoke that may set off a smoke detector inadvertently. Again, evaluate your activity based upon the hierarchy of safety, just as you would for a kitchen experiment.

Many of the considerations used for the kitchen area will be used for the set-up of the outdoor activity. Defining the workspace will be your initial step. Where are you performing the experiment?

Are there any surroundings that may interfere or become a hazard? For example, if you are performing your activity near a garage, are there materials that may be hazardous or could result in an accident? A gasoline can stored in the garage could be a potential source of vapors that may be ignited by your activity. Or if you are performing an experiment that launches a rocket, could this interfere with power lines? You will have to look at your area to ensure that you are not creating additional hazards.

Also, if you decide to perform your experiment in a public place, like a park. You will need to be aware of other individuals that may be come curious about what you are doing. These curiosity seekers could pose additional hazards or may need to be warned of the hazards of the activity you are performing. You may have to define a perimeter of safety, i.e. a buffer of space to minimize potential risks to the observers.

Once you have defined your location and workspace, the considerations are much the same as those used in the kitchen. However, you may also need to take into account wind direction for vapors and smoke. You want to have the observers located away from the potential vapors and/or smoke. Otherwise, you will set up your work area to ensure that the materials as well as the safety materials are easily accessible. You still need to ensure that food and/or drink is not in the work area. And, that proper personal equipment is in use.

You may wish to include barricades and shields depending on the type and size of the experiment and or activity. The key here is to review your activity prior to performing. This help you predict the space and safety requirements needed.

Reviewing the Planned Activity

As previously indicated, reviewing your planned activity is the first step in ensuring a safe activity. You can find dozens of science activity books at your local library and bookstores. A quick search on Amazon.com revealed over 50,000 titles in books. This doesn't even count the number of science activities that may be found on the internet.

However, not all activities are created equal. Some are targeted by age and/or skill. Some are designed to focus on a specific area of science – physics, chemistry, toys, biology, etc. Some are to help with science fairs. Some are for groups and/or classrooms. Depending upon why you have selected to perform hands-on activities, your particular subset may still be filled with a variety of potential activities.

Initial Considerations

As seen above, the numbers of potential activities and sources of these activities may be endless. This makes it a bit harder for you the parent to assess the actual safety of the activity. You may wish to get your activities from a reputable source like the American Chemical Society or the National Science Teachers Association. (There are several source websites listed in the Resource Section of this Book) But, this does not mean that other books and/or activities aren't safe nor does it mean that the activity doesn't have hazards associated with them. You, the parent or responsible adult, will have to do your own due diligence for your particular activity.

Most activity books will start with an introduction and/or guidelines as to how to use the book. If you are getting one-off activities from the internet, there is usually an introduction to the experiment or activity as well. In these introductory materials look for the safety recommendations and/or guidelines. In general, these guidelines state:

- Read the Activity First
- Collect the Needed Supplies
- Perform the Activity or Experiment
- Observe

Many books will also contain the general safety guidelines:

- Have an adult with you or perform under adult supervision
- Be aware of your surroundings
- Keep your area neat and clean
- Put away your materials after use
- Read labels – look for key words such as caution, warning, hazard
- Make labels for your bottles, jars, cups, etc.
- Use proper safety equipment such as goggles
- Do not look directly into the sun
- Pay attention
- ....

The list may go on depending upon your author or developer. While these are a good start, you as the responsible adult, will have to have both your own house safety guidelines, and specific instructions for each and every activity. A set of house safety guidelines that you may use as your starting point have been included in the Code of Conduct at the end of this book.

Once you have established the general safety guidelines, now it is time to start evaluating your particular activity. As indicated, first, read the activity. Most activities follow a standard pattern:

- Introduction and/or purpose
- Materials needed
- Procedure
- Explanation and/or discussion

Others may have additional sections depending upon how the experiments or activities were designed. If they are designed as a formal laboratory experiment, your materials may include a data collection sheet, specific questions, mathematics, etc. Some may even include a safety section, Yippeee! But, if they don't include a safety section, you can create your own following the steps provided here.

Developing the Safety Section

You as the responsible adult may have to develop your own safety section. To help with this, let's look at a standard science activity that we have all done on one or more occasions; the demonstration of a chemical reaction when mixing an acid and a base.

So, here is our sample activity:

Reactions between Acids and Bases

Purpose: To show that acids and bases react.

Materials:
>   Two clear containers
>   Water
>   Baking Soda
>   White Vinegar

What to do:

1. Fill one container about half full with water. Then mix approximately 2 teaspoons of baking soda into the water. Stir until dissolved.

2. Fill the other container about half full with the white vinegar.

3. Now pour the vinegar into the glass with the baking soda solution.

What did you observe?

When the vinegar was mixed with the baking soda solution, you should have seen fizzing and bubbling. The vinegar was reacting with the baking soda in the solution. We typically call vinegar, an acid; and the baking soda solution, a base. Thus, you observed the reaction between an acid and a base.

Reading through this simple activity, now knowing what you know about basic safety; you probably observed several issues. These included but are not limited to:

1) There was no safety discussion at all.
2) There were no recommendations for safety procedures and/or equipment
3) The size of the containers were not specified, thus, the amount of water used was not specified which can impact the size and/or violence of the reaction.

You are going to have to do some modifications to the activity to ensure the safety of the participants. You may even need to test it out to fine tune the experiment before allowing children to perform it. Here are a series of questions that you can ask to help you develop your safety procedures:

1) Are there any hazards associated with the materials being used?
2) What is the anticipated behavior and/or outcome of the activity?
3) What safety equipment should be used based upon the hazards of the materials and the anticipated behavior and/or outcome of the activity?
4) What are the potential bad outcomes from the activity if something goes wrong?

Let's look at these questions in more detail:

Are there any hazards associated with the materials being used?

In this experiment we are familiar with baking soda, water, glass containers, and white vinegar. But, when was the last time you really looked and read the label on baking soda and white vinegar?

A typical box will have one side dedicated to Nutrition Facts, ingredients, and how to contact the producer or manufacturer. The first thing that you might notice is that the ingredient is sodium bicarbonate. Another side of the box, may have helpful tips and/or recipes. You likely to also find a Drug Facts box on baking soda as well. Sodium bicarbonate is used as an antacid. There are specific warnings about the material. One of which is "to avoid serious injury, do not take until the powder is completely dissolved."

A review of the vinegar label provides Nutrition Facts and potential uses. But, recall there are hazards associated with mixing vinegar with the wrong things, the Dangerous Mixtures Sidebar. So, you may have to make your students aware that they cannot just mix vinegar with just anything.

Then there are the physical hazards associated with glass containers if that is what you chose to use for your mixing vessels. There is a physical hazard associated with water, if it should be spilled.

What is the anticipated behavior and/or outcome of the activity?

For this activity, you would anticipate that the reaction is going to produce bubbling and fizzing. But, it is uncertain as to how violent this reaction may be. As this is an activity, you are likely familiar with from previous experience; you may be able to anticipate how it will react. Note: the activity did not mention how fast to mix the two solutions, thus, there is an uncertainty there as well that may impact your safety considerations.

What safety equipment is likely to be needed?

From the previous two responses, you are likely to anticipate the following needs:

• Safety goggles to prevent the solutions from entering the eyes due to the reaction, spilling and or mixing.

• Gloves may need be worn even though these are household materials. Read the labels and other safety information associated with the materials to be used to determine if gloves are recommended. See the box on glove use.

• Spill containment, it is likely that water and/or the solutions will spill during the activity. You will need to make sure that you can clean up the spills right away and/or perform the activity in such a manner as to collect the spills.

What are the potential bad outcomes?

If glass containers are used, it is likely that they may drop and break. You can minimize this hazard by using clear plastic cups.

If the reaction occurs in a closed system, i.e. rather than using cups, your young experiments put the reaction in a jar and close the lid, you may have a resulting explosion. You can minimize this hazard by preventing this from occurring.

Slips due to spillage of water and/or solutions.

Having gone through the analysis, you are ready to develop your own safety section, as well as make modifications to activity to ensure safety. There should be two parts to the safety section, personal protective equipment required and precautions. So, for our activity:

Personal Protective Equipment

o Splash Goggles
o Gloves (vinyl gloves will be sufficient here)
o Science Activity Smock

Precautions

o Never mix vinegar with any other material
   unless directed to do so by your parents
o Do not drink the baking soda solution
o Use care when pouring the liquid
o Clean up spills immediately

In addition, you will want to make notes and modifications to your activity. Your activity may now look something like that shown in the figure.

Now, that you understand the process, here is a checklist to help you develop your safety section. A completed checklist for the acid/base activity has also been provided as an example.

Safety Gear

• Splash Goggles
• Vinyl Gloves
• Science Smock

Precautions

Never mix vinegar with anything unless your parent says it is OK
Do not drink the solutions
Clean up spills right away

Purpose – To show that acids and bases react.

Materials –

Two clear containers	Use two 12 oz clear plastic cups
Water	Will also need a spoon and a measuring teaspoon
Baking Soda	
White Vinegar	Use about 4 ounces of water - to provide room in cups

What to do –

1. Fill one container about half full with water. Then mix approximately 2 teaspoons of baking soda into the water. Stir until dissolved.
2. Fill the other container about half full with the white vinegar.
3. Now pour the vinegar into the glass with the baking soda solution.

Want to pour slowly to minimize the speed of reaction

What did you observe –

When the vinegar was mixed with the baking soda solution, you should have seen fizzing and bubbling. The vinegar was reacting with the baking soda in the solution. We typically call vinegar, an acid; and the baking soda solution, a base. Thus, you observed the reaction between an acid and a base.

Hands-On Activity
Safety Checklist

Are there any hazards associated with the materials being used?

	Material to be Used	Warnings/Hazards Known	Warnings/Hazards Known from Label	Warnings/Hazards Known from Safety Data Sheet or other Source
1.				
2.				
3.				
4.				
5.				

What is the anticipated behavior and/or outcome of the activity?

___ Spills   ___ Fire   ___ Explosion   ___ Heat

___ Cold   ___ Solids   ___ Projectiles   ___ Loud Noise

___ Other: _____

___ Other: _____

What safety equipment should be used based upon the hazards of the materials and the anticipated behavior and/or outcome of the activity?

___ Splash Goggles   ___ Safety Glasses   ___ Gloves – Type: _____

___ Science Smock   ___ Fire Extinguisher   ___ Barricade/Perimeter – Type: _____

___ Spill Clean-up   ___ First Aid Kit   ___ Other: _____

___ Other: _____

What are the potential bad outcomes from the activity if something goes wrong?

_____

_____

_____

_____

Hands-On Activity
Safety Checklist

Are there any hazards associated with the materials being used?

	Material to be Used	Warnings/Hazards Known	Warnings/Hazards Known from Label	Warnings/Hazards Known from Safety Data Sheet or other Source
1.	Baking Soda		Do not Eat	
2.	Vinegar	Slippery when spilled		Do not mix with other chemicals
3.	Water	Slippery when spilled		
4.	Glass	Breaks		
5.				

What is the anticipated behavior and/or outcome of the activity?

X Spills   ___ Fire   ___ Explosion   ___ Heat

___ Cold   ___ Solids   ___ Projectiles   ___ Loud Noise

X Other: _May react violently if mixed to fast, or overflow_

___ Other: _____

What safety equipment should be used based upon the hazards of the materials and the anticipated behavior and/or outcome of the activity?

X Splash Goggles   ___ Safety Glasses   X Gloves – Type: _Vinyl_

X Science Smock   ___ Fire Extinguisher   ___ Barricade/Perimeter – Type: _____

X Spill Clean-up   ___ First Aid Kit   ___ Other: _____

___ Other: _____

What are the potential bad outcomes from the activity if something goes wrong?

Glass breakage - substitute clear plastic

Spills

Do not react in a closed container - don't have closed containers available

_____

Other sources of safety information

While our example focused on something that most people are familiar with, you may run across materials with which you are not familiar. In these cases, you may need to obtain more information than just what is available on the label of the material. Common materials that you may run across in these hands-on activities may include:

- Alum
- Ferric acetate
- Laundry bluing
- Steel wool
- Petroleum Jelly
- Borax
- Hydrogen Peroxide
- Bleach
- Washing Soda
- Tincture of Iodine

First you may be wondering just where to get this stuff, and second what kinds of hazards are associated with each one. Usually, the easiest way to get information about a material is a quick search on the internet. This can help you locate where the material can be purchased and good information about the potential hazards associated with the material.

Let's start with alum. Alum is a hydrated form of potassium aluminum sulfate. You can generally found in the spice section of your supermarket. It is an ingredient used in baking powder and can be found in potash. To find out information about potential hazards, you can do a search with the material name and precautions; or look for a safety data sheet (these have also been called material safety data sheets, but recently the regulations changes to call them safety data sheets).

For alum, you can find fairly quickly that the material is an eye and skin irritant. Thus, should be handled carefully and personal protective equipment should include gloves and eye protection.

In addition to this general search, you can search the manufacturer's website for information. The product label may lead you directly to their website. Baking soda, borax, bleach, and other products have safety and product information on their websites. These are excellent resources for you. In some cases you may be able to find other fun facts, and activities through these sites.

One Last Consideration

There is one last consideration that has yet to be discussed – how to discard used materials. As previously indicated most of the materials that you will be using will be materials that can be readily found at home. But, occasionally you may run into something that is a bit different. Either way, you need to make sure that materials are disposed of properly. Most manufacturers include on the label or in the information that accompanies the product how to properly dispose of them. Even some communities have websites or informational brochures about how to properly handle wastes. Please be sure to review these materials and dispose of your waste materials properly.

Making Safety a Hands-On Activity

Even though you are the responsible adult, you don't have to go it alone in your development of the safety precautions for your activity. You can actually make it part of the activity itself. As you have seen, reading the activity is the first step. So, read the activity together. Ask questions about the materials that will be used. Ask "what-if" questions about doing the experiment out of order. Ask what the students might think will happen.

These questions and insights may provide you with more information about the activity itself, as well as how your students are going to approach the activity. You can have the students read the labels, and/or do the internet search for more information.

Ask them what is unclear about the activity. In our previous example, how much water is going to be used was unclear, as well as how one should make the solution. The students are likely to pick up on these things faster than the adult, because they are coming to the experiment or activity with no previous experience. So, having the students participate in the review is a critical step. It is going to show you additional potential hazards that you may never have considered.

Safety is a team effort. It is not something that can be done in a vacuum. You may be the responsible adult, but all that means is that you have more experience to build upon. All the participants in the hands-on activity have to have a safety mindset. Everyone needs to think through the activity. Everyone needs to be aware of the hazards, as the actions of a single individual can impact everyone else. Safety is a hands-on activity and should not be an afterthought. Safety needs to be included from the very first thought of conducting the experiment.

Hopefully, this brief book will help you prepare for your hands-on activities. The intent is to provide you with tools and resources to make your activities safe and therefore more fun for everyone. As indicated earlier, there is no way to predict all the situations you might encounter. Therefore, you will have to be the responsible party and think through the activity, look for the hazards and prepare for them. Using the checklist and being familiar with the safety hierarchy should take you a long way toward completing this task. Science should be fun. Science should have an element of discovery. But, we don't want this discovery element leading to an injury.

Stay Safe and
Have Fun
Practicing your
Hands-On Science

Resources

Hands-On Activity
Safety Checklist

Are there any hazards associated with the materials being used?

	Material to be Used	Warnings/Hazards Known	Warnings/Hazards Known from Label	Warnings/Hazards Known from Safety Data Sheet or other Source
1.				
2.				
3.				
4.				
5.				

What is the anticipated behavior and/or outcome of the activity?

___ Spills   ___ Fire   ___ Explosion   ___ Heat

___ Cold   ___ Solids   ___ Projectiles   ___ Loud Noise

___ Other: _____

___ Other: _____

What safety equipment should be used based upon the hazards of the materials and the anticipated behavior and/or outcome of the activity?

___ Splash Goggles   ___ Safety Glasses   ___ Gloves – Type: _____

___ Science Smock   ___ Fire Extinguisher   ___ Barricade/Perimeter – Type: _____

___ Spill Clean-Up   ___ First Aid Kit   ___ Other: _____

___ Other: _____

What are the potential bad outcomes from the activity if something goes wrong?

_____

_____

_____

_____

PDF Copies of the Hands-On Safety Checklist can be found at:

http://www.sophicpursuits.com/Science-Education.html

Hands-On Activity
Safety Section

Using the information from the Safety Checklist, create the safety sections

Personal Protective and Safety Equipment:

Personal Protective Equipment:

_____

_____

_____

_____

_____

Safety Equipment:

_____

_____

_____

Precautions:

- _____

_____

- _____

_____

- _____

_____

- _____

_____

- _____

_____

PDF Copies of the Hands-On Safety Section can be found at:

http://www.sophicpursuits.com/Science-Education.html

Safety and Education Resource Websites

American Chemical Society Education -
   http://www.acs.org/content/acs/en/education/resources.html

American Red Cross -
   http://www.redcross.org/
  Types of Emergencies -
   http://www.redcross.org/prepare/disaster
  Tools and Resources -
   http://www.redcross.org/prepare/disaster-safety-library
  Instructor Materials -
   http://www.instructorscorner.org/resourcesforschools/

Center for Disease Control and Prevention -
   http://www.cdc.gov/

National Fire Protection Association -
   http://www.nfpa.org/
  General Safety Information -
     http://www.nfpa.org/safety-information
  Safety Tip Sheets -
   http://www.nfpa.org/safety-information/safety-tip-sheets
  For Children
    Sparky the Fire Dog -
     http://www.nfpa.org/safety-information/sparky-the-fire-dog

National Safety Council -
   http://www.nsc.org/pages/home.aspx

National Science Teachers Association
   http://www.nsta.org/

Ready.gov
   http://www.ready.gov/household-chemical-emergencies

Laboratory and Safety Supplies

(Note the author does not endorse or recommend a specific site. These sites are provided for information only.)

Home Science Tools -
> http://www.hometrainingtools.com/

Lakeshore Learning -
> http://www.lakeshorelearning.com/

OnLineScienceMall -
> http://www.onlinesciencemall.com/

Code of Conduct

Only conduct hands-on activities with the
permission and supervision of an adult.

Use appropriate safety equipment and
wear your safety gear.

Be careful of hot and/or cold surfaces, watch for
sharp edges, and be careful not to break glass.

No horseplay when conducting
hands-on activities.

Follow written and verbal instructions.

Be careful not to contaminate food and
beverages, do not eat or drink while
conducting hands-on activities.

Keep your area neat and tidy.

Dispose of trash and wastes as
directed by an adult.

Clean all materials prior to reusing them.

If you have a question ask the adult
that is helping you.

About the Author

For the past 25 years, Dr. Frankie Wood-Black has been in the petro-chemical and refining industries, but has now turned her energies toward her true passion, providing science education. She is a member of the American Chemical Society as well as several other scientific societies.

But, she has focused her volunteer time toward providing science related resources and materials to schools, parents, and the general public. She has been active with National Chemistry Week, science in the classroom, and chemical health and safety.

As the mother of two children with college degrees, both home schooled through high school, Dr. Wood-Black understands the need for specific hands on curricula for the home school student and parents. She has been through the struggle of finding the right text for the student's learning style, finding the supplies and materials, and the need for documentation of the learning. Thus, her materials are designed not only for the student but also for the one that will be presenting the material.

Frankie received her degree in physics from Oklahoma State University, has an MBA, and is a Registered Environmental Manager. She has seen the public school system through the eyes of a military child, 13 different schools from kindergarten through senior in high school and understands the current education complexities. She has been working with universities, junior colleges, learning centers, and other educational institutions to understand the challenges associated with the new common core requirements. And, she has mentored several new college graduates as they have transitioned from the education system into the working world.

53

www.ingramcontent.com/pod-product-compliance
Lightning Source LLC
Chambersburg PA
CBHW050523210326
41520CB00012B/2413

9 781940 843018